taking away

Author
Karen Bryant-Mole

Wayland

Starting Maths

Numbers
Shapes
Weight
Adding
Measuring
Taking Away

Series Editor: Deborah Elliott
Picture Editor: Zoë Hargreaves

First published in 1990 by
Wayland (Publishers) Limited
61 Western Road, Hove,
East Sussex, BN3 1JD, England

British Library Cataloguing in Publication Data
Bryant-Mole, Karen
Taking Away.
1. Arithmetic. Subtraction
I. Title II. Series
513.2

HARDBACK ISBN 1-85210-876-2

PAPERBACK ISBN 0-7502-0584-9

Typeset by Nicola Taylor, Wayland
Printed by Rotolito Lombarda S.p.A., Italy
Bound by Casterman S.A., Belgium

All the words that appear in **bold** are explained in the glossary on page 30.

Contents

Do you know your numbers?

Look at the boxes and packets opposite. Each has a number on it. Can you work out how many items there should be in each?

Count them and see if you are right.

Here are some cheese boxes. The numbers on the lids tell you how many **portions** of cheese are in each box.

Which group of cheese portions belongs in which box?

Can you put numbers in the right order?

When we count from 1 to 10 we always say the numbers in the same order.

The set of books opposite is numbered from 1 to 10. Can you start at 1 and read through all the numbers to 10?

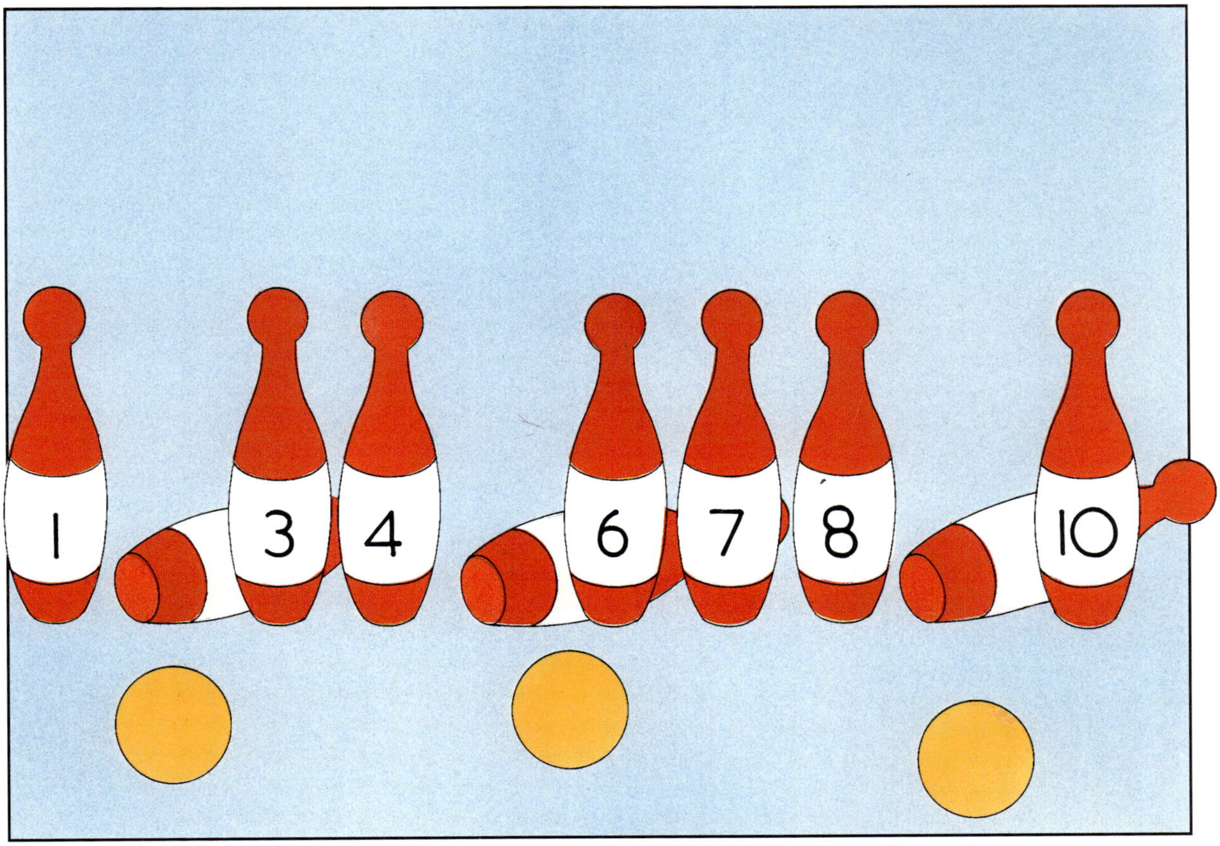

These skittles are numbered from 1 to 10. Some of them have been knocked down and we can't see their numbers. Can you work out the numbers of the skittles that have been knocked down?

Is it 'more' or 'less'?

Sammy and Kate are playing in the sand-pit.

How many buckets can you see?
How many spades?
Is the number of spades more than or less than the number of buckets?

To check the answer, the children match up the buckets and spades by putting a spade in as many buckets as possible.

Is the number of cups less than the number of straws?

If you are not sure, find the same number of cups and straws and try it for yourself.

Can you take 1 away?

How many apples can you see on the plate opposite?

How many do you think there will be if Helene takes 1 away?

Were you right?

Here are some more bowls of fruit.

If 1 piece of fruit is eaten from every bowl, how many will be left in each bowl? (You could cover 1 piece of fruit with your finger if this helps.)

Can you take away more than 1?

How many children are playing on the climbing frame?

If 2 of them decide to go and play on the swings, how many will be left on the climbing frame?
The answer is 4. Were you right?

Pretend that you have 7 books. Which of the groups below shows how many you would have left if you gave 3 to a friend?
Which group shows how many you would have if you gave away 5?
Which shows how many you would have if you gave away 4?

Can you write a sum?

How many cakes are there on the plate?
How many children can you see?

If each child takes a cake, can you work out how many cakes will be left on the plate?
Instead of writing, there are 5 cakes on the plate and then 3 are taken away so there are 2 left, we could write a **sum**.
To write a sum we need to use **symbols** as well as numbers.
This symbol: – means 'take away' or 'subtract'.
This symbol: = means 'makes' or 'equals'.
The sum that tells the story of the children and the cakes is 5 – 3 = 2.

How many bees are on the flowers?

How many fly away?

How many are left?

Which of these sums tells the story of the bees and the flowers?

4 – 2 = 2   6 – 3 = 3   6 – 2 = 4

Can you work out these sums?

David and his mum are buying some wallpaper to decorate David's bedroom. The shop has 10 rolls of the wallpaper that David wants. They need to buy 6 rolls of wallpaper.
How many rolls will the shop have left?

The sum for this story is $10 - 6 = 4$.

Here is a necklace with 10 beads. Use the necklace to help you work out the answers to these sums.

$10 - 3 = ?$ $10 - 5 = ?$

$10 - 8 = ?$ $10 - 2 = ?$

$10 - 1 = ?$ $10 - 7 = ?$

$10 - 9 = ?$ $10 - 4 = ?$

Can you be a magician?

Did you know that you can do magic tricks with taking away sums?

Look at the numbers on each magic card. Do you notice something? They both use the same numbers.

Find some counters and try it for yourself. Remember that the first number in each pair of sums must be the same.

8 – 3 = 5 so 8 – 5 = 3.
4 – 1 = 3 so 4 – 3 = 1.

Use the red flags to work out the answers to the sums on the blue flags.

Can you find the difference?

Ben and Edward both collect badges. Ben has 9 badges and Edward has 12.

They want to find the difference between these two numbers.
They can do this with a subtraction sum.
12 – 9 = 3

Look at these pictures. Count up how many there are in each row of objects and find the difference between the rows.

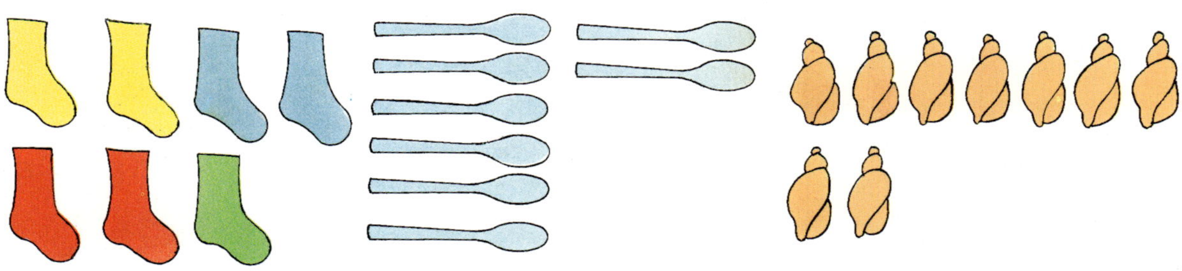

Which of these sums shows the difference between the rows of socks?
Which shows the difference between the rows of spoons?
Which shows the difference between the rows of shells?
6 – 2 = 4   4 – 3 = 1   7 – 2 = 5

Can you count to 20?

Here are two **jigsaws**. Count up the number of pieces in each puzzle.
Which puzzle do you think would be the easiest? Why?

Which set of pieces made this jigsaw?

Which set of pieces made this jigsaw?

These buses should have been parked in the correct order of their numbers from 11 to 20. Two of the drivers got muddled up. Which two buses are in the wrong places?

Can you take away from 20?

Patrick and his dad are buying food for their dog. There are 20 tins of the dog's favourite food. They need 7 tins. How many tins will be left on the shelf?

You could write this as 20 – 7 = 13.

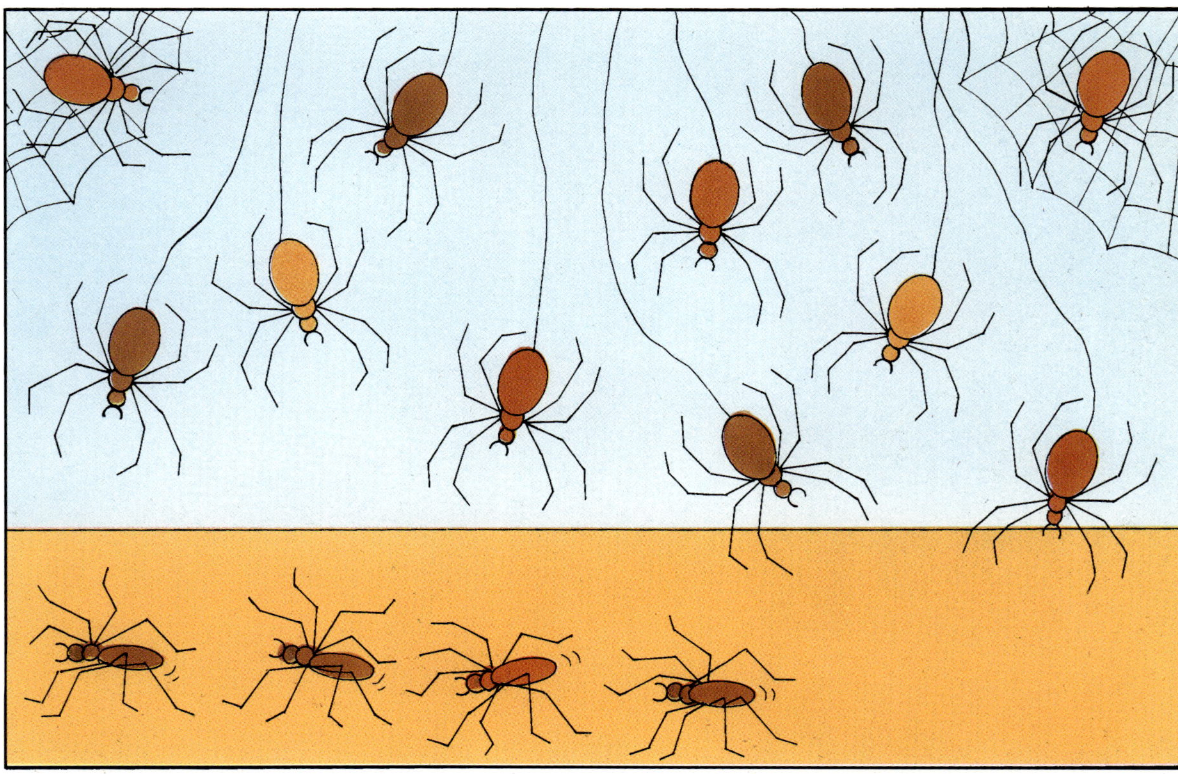

Look at these spiders.
Count how many spiders there are altogether.
How many are running away?
How many spiders will be left?

Can you work out sums?

John and Will are setting the table. There are 4 people in their family so they need 4 forks, 4 knives and 4 spoons.

In the **cutlery** drawer there are 8 forks, 9 knives and 7 spoons.

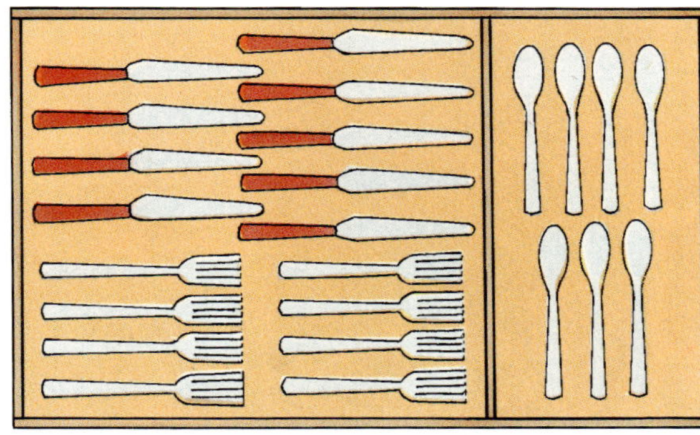

We can use sums to show how many knives, forks and spoons will be left in the drawer after the children have set the table.

The sum for the forks is 8 – 4 = 4.
Can you work out the sums for the knives and spoons?

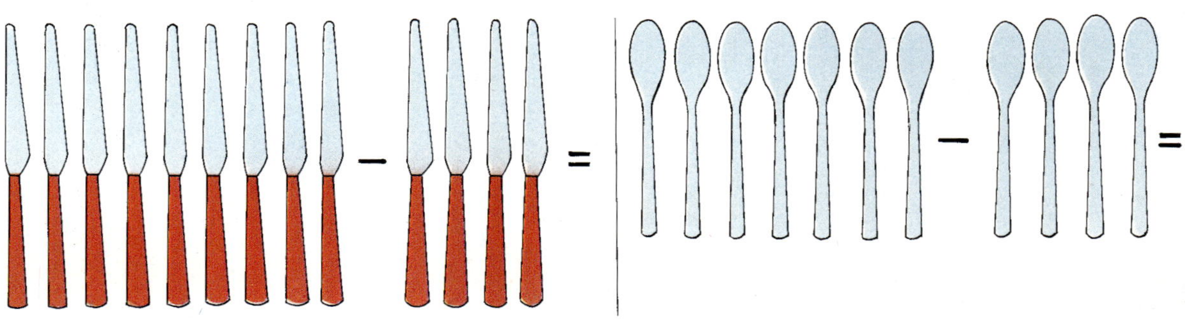

Can you work out what the sum should be?

Annabel and Melanie are playing with building blocks. They have 20 blocks altogether. If they put away the blue blocks the sum that would show how many blocks are left is 20 – 3 = 17.

Can you work out the sum that would show how many blocks would be left if they put away the yellow blocks instead? What about the red blocks? And the green blocks?

How many balloons are there altogether?

Which of these sums shows how many would be left if all the purple balloons popped?
Which shows how many would be left if it was the orange balloons that popped and not the purple balloons?
What if it was the pink balloons that had popped?

16 – 3 = 13 16 – 5 = 11 16 – 8 = 8

Glossary

Cutlery The things we use to eat with, such as knives, forks and spoons.

Jigsaws Puzzles made of odd-shaped pieces that are fitted together to make a picture.

Portions Parts or shares of something, such as food.

Pretend To make believe that something is true.

Sum When two or more things are added together or subtracted (taken away).

Symbols Signs that we use instead of words, for example, – is the symbol for taking away, and + is the symbol for adding.

Books to Read

First Book of Numbers, Angela Wilkes and Claudia Zeff (Usborne, 1982).

Numbers, Henry Pluckrose (Franklin Watts, 1987).

Starting to Subtract, Karen Bryant-Mole and Jenny Tyler (Usborne, 1989).

Picture acknowledgements

The publisher would like to thank the following for supplying the pictures used in this book: Chapel Studios/Zul Mukhida 4, 6, 10, 11, 20, 22, 23; Tim Woodcock 8, 9, 12, 16, 18, 24, 26, 28. All artwork is by John Yates.

Index